by Rachel McConnell

Why you need a content team and how to build one

Introduction	9
Chapter 1 **What is content?**	15
Chapter 2 **What's your current content maturity?**	21
Chapter 3 **Why content at the heart of your products is the holy grail**	33
Chapter 4 **How to use content experts in the digital design process**	39
Chapter 5 **How a content team adds value**	47
Chapter 6 **How should the team be structured?**	53
Chapter 7 **What are the roles you need in your team?**	63
Chapter 8 **Behaviours to look for**	83
Chapter 9 **Recruiting your team**	89
Chapter 10 **Effective team management**	99
Contributors and further reading	111

Introduction

Content – also defined as 'a thing that is held, or included within something'.

It seems simple, and yet somehow it's a word that causes much confusion and controversy within businesses.

In a world before multimedia, content was largely copy (maybe with a few images thrown in). Indeed the word 'contents' is even used to describe the sections that sit within a book, paper, or magazine, so I suppose you can understand the assumption that content always means writing.

The digital revolution brought with it many, many new avenues of communication.

We have social media, video streaming, programmatic advertising, online chat, to name but a few – and virtually all of our 'shopfronts' are now online. We've found more ways than ever to converse with customers.

Content has become a 'catch-all' for a discipline that's as varied as it is misunderstood.

Of course much of our content is still writing – on and offline. But writing has existed for over 5000 years. It's so central to all we do everyday that it's often taken for granted. We're in a situation where organisations are confused about content, and the role that writers and content experts can play in building a successful business.

The digital age brought about the need to move quickly, learn fast and adapt, because technology moves on before you've had time to say 'pick up a pen'. So organisations have looked towards agile tech companies to learn from. They've adopted methods such as 'agile', which allow them to move at pace and react to change, and design-thinking, which allows them to study user behaviour and analytics, and identify opportunities for growth.

New disciplines emerged, such as UX (user experience), and designers and developers forged ahead in creating online journeys and digital porducts. But while design and build teams have obvious roles (the clue's in the name, right?) it can be less obvious just where your content creators fit in. For this reason, many digital teams still don't hire content experts – particularly to help define strategy or to create product content. Content and copy is integral to good design, yet it's often an afterthought, or provided by other disciplines. In some businesses it's created by designers, developers, or even marketing managers, which undermines the value that real content experts can bring to your business.

In my mind, there's no doubt that content and copy form the fundamental building blocks of a user's experience with a brand, particularly online. Good content can be the difference between an average experience, and an amazing one. And good content should flow through the entire customer experience – not just be an isolated section of a web journey. Good content ensures your brand experience hangs together.

The best web content should be so intuitive that a user doesn't even have to think about it – they should just fly through a web journey. But just because it's forgettable, we shouldn't forget that someone has to create it. And that person should be a content specialist.

Despite most of us having been able to write since primary school, and all of us owning pens, pencils and keyboards, it doesn't mean we can all create good digital copy. Often the attitude is that anyone with an English GCSE can turn their hand to copy. Whilst interface design, visual design, and development rely heavily on people skilled to use particular tools, writing copy doesn't, so it's often deemed acceptable that whoever is available at the time 'conjures up some words'.

But as Jean-Luc Godard said:

> "...style is the outside of content, and content the inside of style."

So while anyone can pick up a pen, not everyone can produce clear, concise and compelling content. And that's what you need to bring great design to life.

I'm glad that some digital service companies have started to realise this, and in fact GOV.UK digital service standards now actually states within its criteria for service design that you must have a multi-disciplinary team. It lists the key disciplines as: product manager, service owner, delivery manager, researcher, content designer, designer, and developer. In fact it goes even further to suggest you might need more than one content designer. But crucially, at no point does the ratio between design and content dip below 1:1.

A lot of other companies have been slow to take up the mantle, so in this book I want to challenge organisations to think hard about the role that content currently plays in their business, and understand how much value content experts can bring. When I use the term experts I mean strategists, writers, and producers with experience in their field.

When I speak to other content experts, the challenge we all seem to face is a similar one – even if the role of content is recognised, it's often under-resourced, and is unable to add as much value as it could do to projects. This means design can end up working independently of content.

Martyn Reding is head of user experience at Virgin Atlantic, he says:

> *"Each time I think design for digital products has reached a respectable level of maturity I see a designer working without content and I'm reminded how far we still have to come.*
>
> *I spent the most part of my design career mindlessly chucking lorem ipsum into layouts and getting frustrated when content broke my designs. It was only when I started working with content experts that things fell into place."*

I'd like to start a fundamental shift towards writers being an invaluable part of the design process – so that hand in hand content and design can create beautiful and engaging customer experiences.

In the 1960s, William Bernbach (a co-founder of ad agency DDB) realised the power of design and content working together. Up until then, copywriters and art directors had been working separately, then marrying their work. But he discovered that when they worked together to form a strategy and the creative, the ads were so much more impactful. This way of working moved advertising away from copy-heavy creative to strong visuals with minimal, more concise copy. A lesson many websites could learn from today!

Why wouldn't you want your great visual design to be complemented by great product copy? Sadly, many beautiful websites have been ruined by ambiguous or bad copy. In fact, it's often the smallest details, such as error messaging or instructional text within key user journeys that can be the difference between a fantastic experience or a frustratingly bad one.

Sarah Richards, content strategist, says:

> *"You can spend a lot of time and money on content and if you get it wrong, it can be very costly. I build teams in organisations and often*

> leaders understand bad content can lead to business damage but they don't seem to understand that mediocre and/or confusing content can lead to a huge waste of opportunity. And money. And time. Mediocre content also leads to brand and organisational damage. It's just not as quick to spot."

Getting design teams to collaborate with content experts isn't always easy. I spent three years shifting thinking to this collaborative way of working in one business. But it did change eventually, and in fact shifted much more towards content-led design, which was a huge step-change.

As we move into the future and conversational interfaces become more important — in fact it's predicted that most customer interactions with brands will be via chat — content and copy becomes even more crucial. So companies that haven't already made the shift to content-led design will suffer.

Ellen de Vries, content strategist at Clearleft says:

> "With conversational interfaces, and the ubiquity of interfaces, there is ONLY content. Organisations are having to work more and more with content and domain models, so that they can prepare their content in advance, at source, in the Lego block forms they need (the Create Once Publish Everywhere idea - COPE). Keeping consistency and delivering good quality content across a variety of interfaces is becoming a larger and larger task, so content experts can already begin helping clients prepare for this."

Based on my experiences, I wanted to share my thoughts, some tips for creating a content team, and offer some helpful advice on finding the right content experts to help grow your business and change your ways of working. For content that works effectively — across every channel.

Chapter 1

What is content?

Content is any information, images or media that's available to a user. When it's digital content, this just means it's available on a digital platform. Content can take so many forms, but here are a few examples:

- Navigation labels
- Micro-copy within UI (including chat or voice UI)
- Meta data
- App release notes
- Video
- Images or icons
- Infographics
- Forms
- Proposition headlines
- Social media posts
- Podcasts
- Letters
- Direct mail
- Emails
- Articles
- Adverts

Determining exactly what form your content takes within a customer journey forms the majority of a content expert's role. This is just one of the many benefits a content expert will bring.

The format will depend on who your users are and how they behave, and of course the business objective. Content experts spend a lot of time understanding this, and they'll know how best to present content to a user to get the response required. The goal is the sweet spot where user needs and business needs are met.

The type of content will also depend on the company's brand and communication strategy. For example, if your communication strategy is about relationship building, then mass awareness formats such as display ads might not be appropriate, but email communication might be.

Many companies already employ experts to create their marketing and communication strategies and employ experts to write ad campaigns, customer emails, and their blog posts. But when it comes to the rest of their content – particularly within a digital product or on websites and sign up journeys – they forget that they need specialists.

These days, websites and online journeys are at the heart of your customers' experience. And your brand values need to live at every point of their journey through your content. This means content such as micro-copy and transactional copy should have just as much thought behind it as your advertising campaigns. If not more so.

The brands that value content and do it really well – Innocent, Airbnb, Red Bull, MailChimp, Ocado, Boden, Birchbox are just a few brands that spring to mind – all have content teams full of experts who know exactly how to resonate with their audiences.

They also have clear strategies for each element of content - they know how it all hangs together, and it's consistent, so whether you're looking at a text message, Instagram post, the website, or a physical product, it feels as if it's come from the same brand.

I'm of the opinion that any kind of customer-facing content should be created by experts, who understand your customers, your physical and digital products, your brand strategy, your business objectives, and the end-to-end customer journey.

If you're just asking someone to 'craft some clever words' on an ad hoc basis, without thinking of the bigger picture, you're undervaluing the impact of content.

When it comes to functional content and navigational content, using experts who understand usability can transform your users' experience. Good instructional content can be the difference between making a sale, or a frustrated user leaving your site or app, never to return.

Sara Culver leads the content design team at Slack. She says:

> "Our brand has always been a competitive advantage – folks love the brand voice, which really is pretty different from a lot of B2B and Enterprise software companies – and my team is the steward for that voice in the product. We try to make the experience of using Slack feel very courteous and human, to make it clear that someone has put a lot of thought and care into whatever you're seeing, even (or especially) bits of UI that don't usually get that level of attention from professional writers, like error messages or settings and preferences. People really seem to love the more overtly playful bits of content in the product, and we do as well, but we've worked hard to make

sure the less flashy words are still clear, concise, and human. I think (hope!) this means that the experience of using Slack is simpler and more pleasant than it otherwise would be."

If you're only using content experts to create customer marketing communications, then it might be time for a re-think, so read on!

Chapter 2

What's your current content maturity?

There are many maturity models that have been developed over the years, and if you work in a digital team you may be familiar with the Jacob Nielsen UX maturity model.

I decided to create my own for the modern content age. And whilst it may seem a bit tongue in cheek, it's designed to assess your own organisational capability and identify your gaps.

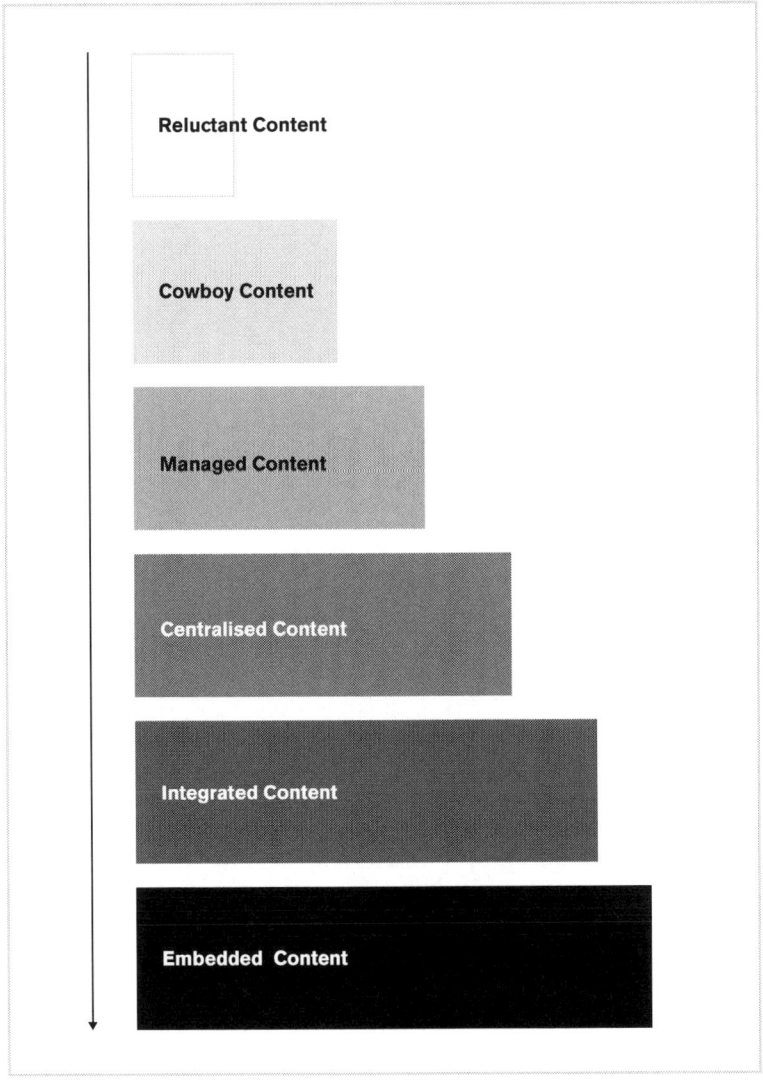

Maturity level one - Reluctant content

I thought about naming this 'unenlightened' but that doesn't quite illustrate the level of resistance that exists in these organisations.

This level can be identified by the following:

Attitude
Anyone can do it can't they?

Resources
Minimal. There might be one or two people who 'like to be creative' but have no specialist content skills.

Production
It's ad hoc and produced reactively, usually responding to a specific need. There's no strategy and no centralised content function.

This could be because the company doesn't feel like they have much content to produce. But what they haven't realised is that content is all the 'stuff' they have on their website. Any words, images, call to actions - it's all content.

It's likely the content that exists hasn't been audited, it's out of date, and has possibly spiralled out of control. For example FAQs might have just become a 'dumping ground' for content. Nothing produced has been based on analysis or research.

For this business, education in the role that content plays here is vital. This company needs to move away from the belief that content is 'just articles', and something they don't really have a need for. Everyone has a need for usable, accessible content.

A content expert going into this business is going to inherit a lot of content debt. Debt is the content that's built up over time and now needs to be cleaned up. The first task they should do is to carry out a content audit of existing content to identify issues (such as out of date or redundant content), and understand how it's been structured. They should look at some analysis or quantitative data to see which content is being viewed, and if possible undertake some qualitative research to see if what's there is meeting user needs. They should also start to implement a publication and maintenance process, which will add in some control and governance.

Any new content requests should be briefed in with an objective, and clear outline of why the content is necessary. If it's not meeting a user need or essential to comply with a business regulation, then it should be put on hold until the content audit has been carried out. It can then be revisited once the site re-structure (if it's required) is done.

Maturity level two - Cowboy content

This company has recognised the need for content, but still doesn't see it as a core function. There might be hostility towards those trying to advocate the creation of a specific team.

This level can be identified by the following:

Attitude
Isn't that what we have agencies for?

Resources
Limited. There might be one or two content managers but often content is produced by freelancers or agencies.

Production
It's reactive to business demands, and tactical, not strategic. Some processes might exist but they're disconnected and not governed, for example there might be one approval process if a content manager creates something, but another when the agency does.

As content is often produced by different teams, such as marketing, design or development teams, it can feel inconsistent.

And when there's a dependency on agencies to create it, content can be seen as expensive asset, or a 'nice to have', so it won't be prioritised above other projects, or worse, shortcuts might be taken.

The problem with this approach is that your content, particularly in user journeys, can feel disingenuous and inconsistent. When the content team is so small and not in control of all the content that's created, they might not have had time to produce loud and clear brand tone of voice guidelines or principles, so this inconsistency will continue.

John Saito is a UX writer at Dropbox. He says:
> "Anyone can write. But a good UX writer knows how to make a product have just enough personality without getting in the way. And a good UX writer knows how to say just enough to help people know what's going on.
>
> I think a good analogy for this is acting. I mean, theoretically, anyone can pretend to be someone else. But if an actor does something that seems out of character, your audience will immediately notice and it

> breaks the scene. It's the same way with writing. If your user suddenly reads a line of text that sounds unlike anything else in your product, it breaks the experience and makes your user start doubting you.
>
> By having a content expert own the words, you can make sure your users keep believing in you."

A UX writer or content designer could be a great next asset to add to a team of content managers, to start instilling this consistency.

The other crucial thing with a company at this stage is to set targets for the internal team that start demonstrating the value of clear consistent content, particularly for in-house work. This will help the team start to show how cost-effective they are, so that there's less of a need to outsource.

A UX writer or content designer can also begin highlighting the issues that can arise with inconsistent content, particularly if a user is moving from one channel to another. For example if a welcome email describes the log in process differently to how it actually works online, you're creating a bad user experience.

It's also a good time to start documenting roles and responsibilities for the team – this will highlight skill gaps, and identify what the creation and publication process could look like to help reduce silos and ensure there's a consistent 'path to live'.

Anyone coming in to manage this content will also inherit the content debt that's built up – disjointed, out of date content, probably not structured in a way that's conducive to a good customer experience. This company has fallen into bad habits which will need breaking!

Maturity level three - Managed content

Now there are one or two people in the business who are content advocates and driving the understanding of content into the rest of the business, but things are still shaky.

Results are coming in, and there's a realisation that content can add value so there may be some investment in training to drive capability. There might be some resource funding too, to help grow the team.

This level can be identified by the following:

Attitude
This is working, let's get more people!

Resources
The basics – a strategist and one or two writers exist.

Production
Business is starting to realise the value of content in user-centred design so has a content designer who works in collaboration with design and research teams, but usability-testing content may still be a little flimsy, and it's still not seen as a fundamental part of the design process.

Strategy, and guidelines are emerging but ongoing governance and measurement needs improvement.

This is a good start, for sure. But this company may still not be thinking about how all their content 'hangs together' across their channels.

They might now have tone of voice guidelines, or an agreed process for approvals. If they do that's great, but once their content is live they may not be measuring it to see whether it's commercially benefiting them, and there's probably a whole heap of content that's live on their site and needs optimising. Having a small team means optimisation often isn't done by a content expert, and again that drives inconsistency.

This content team can probably see lots of opportunities to use content for growth – if only they had the resource to get to it!

Having an under-resourced team also provides a lack of growth opportunities for team members – they're too busy getting the day to day work done to attend training and they might not have a strong team lead driving team growth. This can lead to frustration or boredom and make it hard to retain talent.

In this business, not only does the content team need to grow, it needs to plug skill gaps. It must ensure a long term strategy is created, and that optimisation work is done with a content expert's involvement. Seeing commercial results through content tweaks can help create a 'best practice' content playbook or guidance document that feeds into a longer term strategy. This also helps to demonstrate the difference that good content can make. Once the rest of the business starts seeing these results, they can understand how they'd benefit from having a content expert on every project team.

Maturity level four - Centralised content

The organisation now has a centralised content team and has recognised the need for a strategic approach to content to create a great customer experience.

This level can be identified by:

Attitude
Content is kind of a big deal!

Resources
Strategists, content designers, writers and content managers

Production
There are brand tone of voice guidelines and there's an overarching content strategy. Content strategy is now moving into product development, and is user-centred, although it still may not always be seen as playing a fundamental role in how a product can be developed.

In this business content experts work to agreed customer experience principles, and work with each other, not in silos. Digital scrum teams always have a content designer in place who collaborates with the rest of the team and knows the value of usability-testing. The design might not always be content-first, but it's definitely done in tandem.

This content team has a clear understanding of 'best-practice' and sets standards and key metrics for any piece of work. If content doesn't meet the targets set it will be optimised.

The wider business still probably doesn't know when (and if) they should involve a content expert in projects, and they might be unclear about how influential content can be in shaping the product itself.

A content team in this business needs to start educating the wider business, perhaps by running workshops or pair-writing sessions, and sharing their content successes. They could also do this by taking stakeholders to usability testing sessions, or sharing content-driven optimisation successes. Once you have a few ambassadors in the business who understand the role content plays across the customer's experience of a brand, it will help build momentum.

Maturity level five - Integrated content

Now the business has a fully-resourced, centralised content team and has started to move experts out to act as consultants for the rest of the business.

This level can be identified by the following:

Attitude
Good content is at the heart of all we do.

Resources
Strategists, content designers, writers and content managers who can turn their hand to help other teams and are true content ambassadors.

Production
It's not just about production at this stage, it's about longer term thinking – and the belief that content really can shape a product or change the way it's sold.

Content is produced with re-use in mind but it's also used to drive innovation. Design decisions are documented, learnings are shared across teams. Users are at the heart of content design.

The business has a clear understanding of 'best-practice' and sets standards and key metrics. It wants to scale up the role of content in the design process, and empower the business to employ more content experts where needed.

This is where the wider business can start to feel empowered to know when to enlist content experts for their projects, and when they do, to get them in early on.

In this business it's easy to suggest digital product changes that rely heavily on content – for example shifting towards a chat-led sales model, or voice UI, because the capability and skills exist in-house to do this. In this business it's easy to get swept up in wanting to do and test everything, but your resource might still not be enough to meet demand.

Maturity level six - Embedded content

At this stage, because content is so pivotal to everything the company does, while there might be a head of content or a chief content officer, there's no longer a ring-fenced content team.

This level can be identified by:

Attitude
Content team? What content team?

Resources
Content designers, managers, writers and strategists sit within in the business and all work to the same strategy. They might come together as a community of practice.

Departments have the knowledge and tools to hire their own content experts as and when they need them.

Production
Strategic, joined-up and process-driven. Content is always measured and optimised, and the business invests heavily in building capability. All areas of the business will collaborate with content experts to create the best possible customer experience.

This might seem like the best set up, but the trick here is to ensure consistency and communities of practice are maintained. Otherwise parts of the business could start to go rogue and do their own thing! The strategy needs to also incorporate how to maintain and evolve best practices as content experts around the business learn and adapt, and who has overall responsibility to keep the community together and drive the strategy forward.

Chapter 3

Why content at the heart of your products is the holy grail

Content-led product development

As we move into the future an entire business should be digital (it will be the only way of doing things). Start-ups are digital by their very nature and don't have 'digital' teams, but we're in a weird state of flux where older companies are struggling to shift to a digital model and often have centralised digital teams.

But companies that aren't digital by their nature and who haven't yet adopted a digital first approach often still regard digital as a 'channel' (as it once was when it was just a small part of your marketing mix). Digital is quickly becoming the main route (and in some cases the only route) to your customer. As I said before, if you run a service or sales business, it's likely your website is the heart of your customer journey's ecosystem.

The problem with having a digital team inside a company, is that digital can be thought of by those outside the team as a 'channel', and one that can often the last thought by non-digital product teams, who have already created a product and the messaging without as much customer insight as they could have had.

It means the digital team becomes a delivery service – responding to other teams in the business who want to create digital content but haven't started with the end user in mind – rather than allowing the the digital team to lead the service design.

Good digital teams use design thinking. This is user-centred design process that enables them to develop products and services based on evidence and user insights (the most effective companies are those whose products are designed to meet users' needs).

You can see the conflict that may arise when digital teams are just responsive, and not part of the initial product development.

So much can be gained from engaging your digital teams early. Being given collateral and messages that need to be communicated, without the ability to help mould them into a format better suited to your customer (see my above section about deciding the right format) can feel very frustrating. But it also means the digital team aren't adding as much value as they can to product and service design.

Digital content experts who understand users and their digital behaviour, can add a lot of value at the product development stage (along with research and design). If you're in a non-digital team, why not explore how you could work alongside a multi-disciplinary team from the digital department to see how much it could help enrich your product?

And if your product is purely a digital one such as software, it's even more crucial to have content experts onboard – your content can really make or break the usability of your product.

To end up with a brand, product or service (and the messaging), that's relevant and useful to your customers and doesn't need to be re-written to work better online (or in your offline communications), engage your content experts as early as possible on any project. If you're doing any research, it's particularly valuable for your content experts to hear first-hand from the end users. They'll listen for insights but also listen for the language they use – particular words and phrases – that can help with their content development later on.

Content experts are needed at each stage

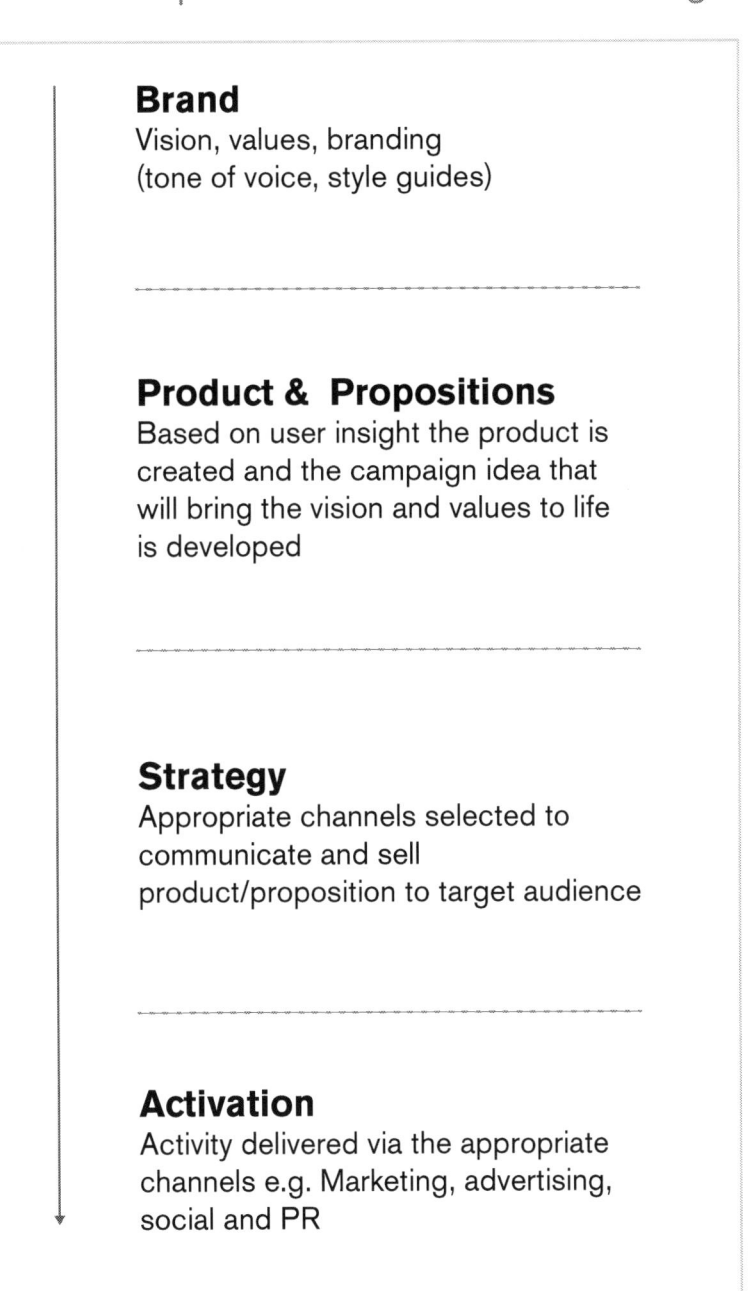

Brand
Vision, values, branding
(tone of voice, style guides)

Product & Propositions
Based on user insight the product is created and the campaign idea that will bring the vision and values to life is developed

Strategy
Appropriate channels selected to communicate and sell product/proposition to target audience

Activation
Activity delivered via the appropriate channels e.g. Marketing, advertising, social and PR

Chapter 4

How to use content experts in the digital design process

You may be lucky enough to work in a fully digital company who already understands the process of design thinking. However that doesn't always mean that content experts will be engaged at the right stage of the process.

I've worked with companies where content designers or UX writers exist, but are brought in far too late on a project to just 'supply the words'. This presents all sorts of issues. Here are the main ones:

Issue 1 - Project delays

When you read about principles for good content production, you'll notice that context is generally right up there. Understanding why you're doing what you're doing, and for who, it pretty vital when creating content. A content designer coming onto a project halfway through will need to ask a hell of a lot of questions to be brought up to speed and may even have to pick their way back through the research that was carried out at the start of the project. They will need to understand the customer insights, and why the design has been done the way it has.

In addition, getting content and copy approved by the relevant stakeholders is time consuming – this process should be started as early as possible on a project to avoid delays towards the end.

Issue 2 - Inaccurate usability testing

Placeholder copy or lorum ipsem in a prototype isn't ever going to give you an accurate read of whether your users understand what they're doing, or buying. We take what we see on screen at face-value, so users will get hung up on words whether you want them to or not. They don't see a difference between form and function, they see a holistic view of a page, so it's also unnatural to test without realistic copy. Content designers or UX writers should also be at usability testing – that way they understand exactly what does, or doesn't work for users, and can iterate accordingly.

Issue 3 - Content and design don't complement each other

When content specialists are brought in late, they might be restricted by designs that don't fit their content strategy. This can lead to disagreements between designers and content designers, who have opposing views on how the information should be presented. It can lead to resentment if content has to be compromised, or designs have to be redone. This conflict will also cause delays.

Issue 4 - Inaccessible content

Content designers understand the importance of accessible content. We cannot assume users have English as their first language, high literacy levels, or don't suffer from any kind of reading difficulty. And just inserting content supplied by marketing or legal teams could mean that jargon and unwieldy sentences catch users out.

Issue 5 - Content is devalued

When you take away the colours, images and interactions of a web page, or app, what are you left with? The words, or the conversation you have with a customer, are the fundamental building blocks of digital web journeys.

Jef Raskin from Apple knew this, and when he designed the UI for the Apple II, he started with pen and paper, and simply wrote the conversation he wanted to have with the user.

Assuming that content and copy can be provided later on in a project undermines this approach, and has the knock-on effect of content teams being seen as of less importance to design as design and research teams.

If you've ever used Jef Raskin's method to create a product journey, you'll understand how powerful it can be. Having taken this approach on some recent product journeys and seen the impact – I'd go as far as saying it can even lead to you changing the product itself.

By starting with the conversation, your designers will have the building blocks to work from.

Issue 6 - Conversion suffers

Message hierarchy, copy, call to action labels, and navigation labels – getting these wrong can harm your conversion. Content designers will ensure your content is organised and produced in a way that makes your website more human, usable and engaging, optimising your sales and keeping your customers coming back for more.

Here's where (and how) I believe experts add value in the process:

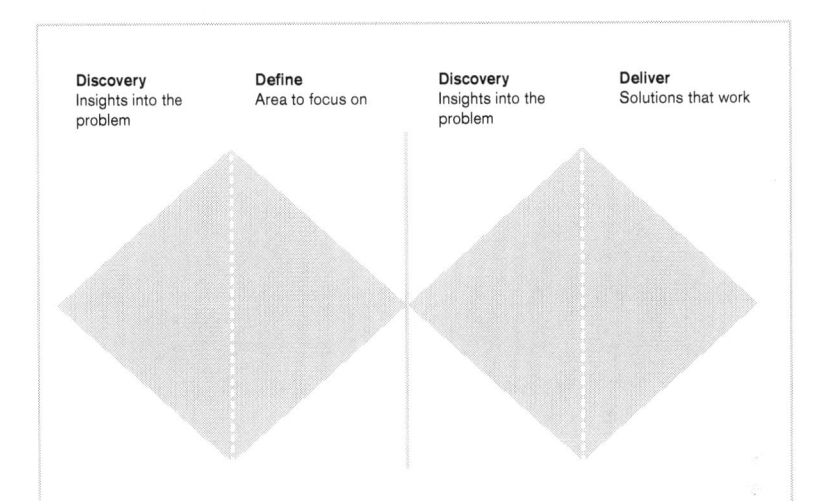

Discovery	Define	Discovery	Deliver
Insights into the problem	Area to focus on	Insights into the problem	Solutions that work

Discovery
A content designer (or UX writer or strategist) should be involved to help understand insights, assist in research by helping to write questions, capture responses and grouping output from research sessions, as well as listen to the language your users use (which will help them later on).

Define
The content designer helps to define the scope, resource, and time needed to deliver the work.

Discovery
The content designer works with a designer to create the user conversation for your product and help to turn this into a prototype. Testing also happens in this phase and during or after testing the content designer can iterate accordingly.

Deliver
Delivering the solution to the build team means that in this phase of a project your content needs to be ready — approved by relevant stakeholders, and documented for the build team in a way that's appropriate (this could be a wireframe or prototype, spreadsheet or via an online tool).

Collaboration is powerful

The beauty of content and design collaborating, is that they may decide together that an image or diagram is more appropriate than words. A content expert is never afraid to shy away from words. In fact the fewer the better. Creating an approach together means that any designs are appropriate for the content, and vice versa. Remember by earlier point about how this approach revolutionised ad creation in the 1960s?

When content and design experts collaborate, you'll also get the optimal user journey, as this can impact functional copy, button labels, or even how data is captured from users.

Once a designer has worked with content experts, they'll quickly become advocates. Martyn Reding says:

> "It wasn't until I started pairing the design process with writing and content strategy that things fell in to place. For me I could seeing how much richer and more effective any solution is when you bring content and design together, from early concepts right through to research and into build. If you're designing a product, a system or a basic static website your task is to design a delivery method for content. If you don't understand and respect that content then you're not a designer, you're just colouring in boxes.
>
> These days I instinctively dismiss any designs that are brought to me with poorly thought-out content design. Heaven help the designer that asks me to review a UI that contains Latin placeholder copy!"

When capturing data from users, a content expert can help determine the best way to capture it. It might be a traditional form, or conversational form. It might be a chatbot or live chat service. It could even be voice UI. Each comes with it's own challenges and benefits, but the expert would work closely with product managers, researchers and designers to decide which option to prototype and test. This is where your expert really brings value, as you'll be testing with accurate and realistic copy right from the start of a project. Content designers also understand how to consider copy for small devices, and when to use techniques such as progressive disclosure to reveal information bit by bit (as, and when the user needs it), or how to layout a page to communicate message hierarchy.

In less mature companies, article writers are often drafted in to create copy for this type of user journey. While I'm not suggesting traditional writers can't adapt, it's a very different skill. In my chapter on roles within a content team, I'll present some more evidence to support this thinking.

Content experts will follow key principles such as brevity, plain English, and short sentences. Having a content lead or head of content in place to set out these principles will ensure a consistent approach across the business.

Some of the opportunities companies that don't use content experts are missing out on are:

- Optimising sales or conversions through simple content tweaks
- Consistency of brand tone of voice across all customer touchpoints including web
- Usable and accessible copy that's designed to get customers to their tasks as quickly as possible
- Logical navigation and clear call to actions on site to drive sales or online servicing
- Engaging, helpful content that resonates with the target audience

Chapter 5

How a content team adds value

There are so many roles your content team play in your business, and that's what brings the value and efficiency to your projects. Here are a few of the things I expect a content team to do:

Be the connectors

Your content team have to work closely with everyone: product owners, document producers, designers, researchers, compliance teams, marketing, PR, developers, and business analysts. They need to understand exactly what they're producing and why, and turn it into something the production team can understand. Sometimes content is self-managed in a content management system (CMS), other times it might need to be produced by a third party or documented within a prototype. Whatever the review, production or publishing process, your content team should facilitate this, so they'll need to build good relationships with all the relevant stakeholders.

Create editorial & communication strategy

As a very minimum your content team need to understand the topics that need to be covered and decide when, where and how to deliver the content (and the best format for usability). The content might be driven by a business need, or by a user need – your goal is content that meets both! A content team will come to understand user needs through research (which may or may not take place alongside researchers and designers – it depends on your business model). If the channel of communication isn't already determined, the team may also be responsible for identifying the most appropriate channels, and creating content to suit those channels. You content team will know how the success (or failure) of the content will be measured, and review and adapt accordingly.

Content experts take time to understand the purpose of their work before they start it, which might mean asking a lot of questions to ensure they've understood the brief. A strategic approach means thinking wider or longer term than just the immediate task in front of them.

Design content

A content team should decide how best to structure content so its easily found and navigated – at a broad site level (information architecture) and also at page level. Perhaps sometimes also at micro-level if they're looking at granular things like meta-data and alt-tags to make sure content is easy to find from search engines. This also links back to the strategy piece; how you want your brand to be seen by potential customers could define what your SEO (Search Engine Optimisation) descriptions say.

Projects such as web redesigns and the design of new user journeys should always have a content expert on the project team. Content is such a fundamental part of the user experience that when it comes to information architecture (IA), page design and journey improvements, a content designer can add so much value – both in terms of how the content is structured, but also the actual copy you use. When it comes to interaction design, they'll bring the perspective of the conversation you should be having with your user.

Content designers can also help determine how a CMS should be engineered to deliver content that reflects the designs and IA, and can work alongside a dev team to ensure a CMS is built as required, and test it (to highlight bugs and defects).

Writing

At the very least you need writers, whose job it is to know who your users are, and what you want them to think, feel or do next. They'll be the guys that create the functional interaction copy (or voice) that's useful and usable for your audience, and accessible for the majority. This is closely linked to content design, but I've separated this as a role that content designers play within their wider remit. See more about the differences later.

Create content workflow & governance processes

Determining how content is created, approved by stakeholders, managed, maintained and monitored, is as important as the content creation itself. Once the content is live someone needs to maintain, update and optimise it. It might quickly become out of date (and in a worst-case scenario, detrimental to a customer) if it's not reviewed and updated regularly. Content teams should put processes in place to check work before publication, to govern the publication process itself, and to review it once live.

Analyse and optimise

Content specialists can determine key performance indicators (KPIs) to measure against, to make sure their content is achieving the objective set. They'll know the difference between dwell time and bounce rates, click-through rates and page views. They should review metrics, and if it's not successful try to ascertain why not, so that they can optimise it. This might be done on a regular basis through A/B testing if they're within a team who purely focus on optimisation. If not, the content team should set clear review points in a year to ensure their content is still appropriate and performing as planned. This could be done through a content audit, which a content expert will know how to do, or through another process.

Be ambassadors

I'm of the view that good content experts want to help others understand digital content principles and help guide all areas of a business to create better content. Content experts have to work with so many people – product owners, marketing teams, subject matter experts, the list goes on. It really makes collaboration less challenging when everyone understands the value of user-centred design. Your content team can share their expertise more widely and build knowledge in multiple ways. This could be through 'lunch and learn' sessions, or on a one-to-one level such as pair-writing (where you help a non-writer to write).

If these seem like valuable assets to your business, but you can see you don't currently have content experts fulfilling these roles, then you've identified some easy opportunities to improve your content capability.

Chapter 6

How should the team be structured?

Before deciding the team you need to have in place, you need to consider the work that needs to be delivered and the skills best place to deliver that work.

Identify who you already have available and what skills they have. Where are the gaps?

Your structure will depend on many things, including:

- Budget
- The wider team and company structure
- Any future projects that need to be resourced and the nature of them
- What's currently done internally and what's done via agencies (could you shift this model easily?)
- How existing content is managed and optimised
- The people (and skills) available to you
- Whether or not your company tends to resource short-term projects with contractors

A quick note about in-house vs contractors

If you're limited on recruiting in-house, there's definitely value in using contractors. But I believe the design work on a new project should be created by your in-house team and not contractors. An in-house content specialist will have a long-term view of what they're creating and leave it in a place where it can be easily picked up by someone else. They already have relationships with internal stakeholders, and know your target audience and customers inside out. They'll take more care over the longevity of the outcome, because they'll have to live with the results and manage whatever they create!

They won't just be interested in how it looks in their portfolio, they'll want to understand how it works functionally.

In-house teams know your business well, and understand how to navigate it to approach and influence subject-matter experts and the key approvers.

Also bear in mind that contractors can leave a company with very little notice, leaving you to have to fill their role. This can lead to delays to projects, so think about which projects should and shouldn't be resourced by contractors.

That said, sometimes contractors can bring an external perspective and 'fresh-eyes'.

I'm in favour if necessary, for ad hoc content demands and supplementary resource can be contracted out, as well as some optimisation work. In fact, great contractors also often help to broaden the skills of an in-house team. If you do use contractors be sure to learn what you can from them before their contract's up, they often bring a wealth of knowledge from their experience with other companies.

A bit about structure

How you resource your team also really depends on your wider team structure.

Flat structures

A flat structure refers to multiple product or scrum teams with equal levels of skill. There may be a central manager or director overseeing the teams, but within the teams there's little hierarchy.

When agile digital teams have scrum teams they should ideally have dedicated content resource on each team to collaborate with research, design, and analytics. The work each scrum team is doing will determine the particular skills you need to meet. You might need a content designer, or a content manager, or a content strategist. In this structure there may not be much opportunity for individuals to progress so it's worth considering this when you recruit. It wouldn't be in your best interest to recruit someone who's expecting fast career progression upwards. They could quickly become frustrated and leave. If you're looking for one person to work across multiple scrum teams, they'll need to have as broad a skill set as possible, which may be hard to find. It's also likely someone with this kind of experience will be looking for a level of seniority which they may not gain, or will be looking for a higher salary.

Hierarchical structures

If your company has senior positions and junior positions you will probably want the management roles to have a broader skill set in order to mentor the junior team members. In this instance, I would probably be looking for the senior member to be a generalist but with more of a leaning towards strategy. Recruit your more senior roles first, then your managers can help determine what they want juniors to be responsible for, and what skills they need. It's also important when hiring, to identify someone who could be a 'second in command' for the team lead, to cover holidays or absence, and generally support them in their work.

Split functions

You might need content team members to sit within different teams – for example brand and marketing, digital, or customer documentation teams. Each of these areas needs a different skill set. If you have the budget you should opt for strategists in brand and email marketing roles, and content designers or UX writers for digital. But if you need people to each sit across one or more of these areas I'd opt for generalists with broader skill sets. In this structure, it becomes even more important to have a team lead – someone who can bring the team members together regularly. This helps ensure that even if they're not sitting or working together on a day to day basis, they still have sight of what they're all working on. This can help increase learning, avoid duplication of effort, and maintain consistency across the different customer touchpoints. Sarah Richards, strategist and digital content consultant, says teams should use each other for "Sanity checking and inspiration! In our content, in our interactions with others and in our industry."

Generalists vs specialists

Generalists

A generalist can turn their hand to almost any kind of content work. They'll know enough about everything but may or may not be an expert in any one particular area. The benefit is they have a broad understanding and and can look at content holistically. On the down side they might need to call upon experts if they're asked to do something beyond their knowledge (for example, I'd class myself as a generalist, so when it comes to SEO, I'd definitely have to consult an expert!).

Generalists can make good team managers if they have specialists working underneath them. A manager needs to see the bigger picture and be able to identify when specialists skills are needed.

Specialists

A specialist will prefer to focus on one particular area of expertise and grow their knowledge in that area. Sometimes it can be tricky to ask a specialist to do something outside of their comfort zone or do work they're not comfortable with. Think carefully about where you need specialists and whether these demands will shift. If you need a very specialist role for just a short amount of time, perhaps you should be looking for a short-term contractor.

Your first hire

Let's assume you don't have a content team at all, and you've got people in your business all chipping in to produce content.

What you need is a generalist who can turn their hand to strategy, and create process, but also someone who has writing and usability skills that can design content within your user journeys.

This person will have their work cut out. They'll be starting small but aiming big. As they start to demonstrate the value of content, and create a longer term view, you should start to invest in broadening the team, and adding specialists in the areas you need them. Ensure your first hire is experienced enough to take on the whole company as the sole content expert, and ambitious enough to grow their role. It's a tough job. They'll have to tread carefully to change behaviour, instil discipline and rigour, and influence reluctant stakeholders.

As the team grows this person will need to be able to adapt from a hands-on doer, into a manager, so look for someone with people skills too!

Structure examples

Your structure will be determined by your business needs, which may be unique to you, but here are some examples of how you could shape your content team if you needed content for brand, marketing, and product teams. I'm using examples which include hierarchy so there's room to grow junior team members. They also show how you could scale by adding more editorial writers or content designers as needed. It might be even be that your business needs more content managers, depending on how many sites you have to manage.

Managed

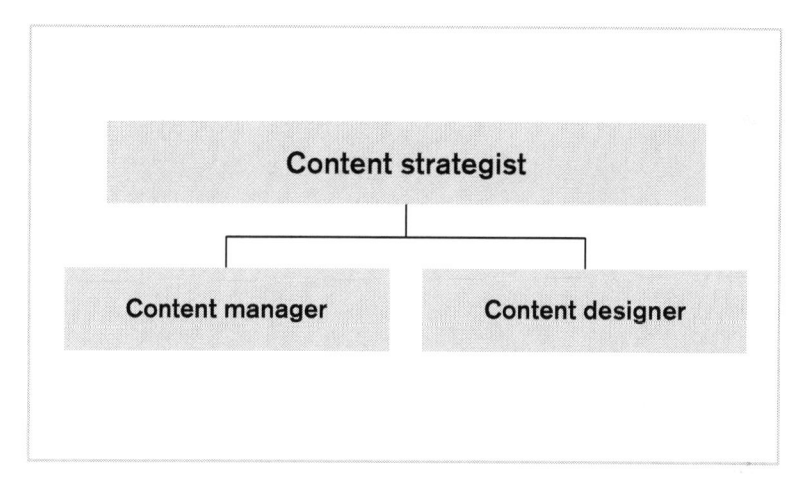

```
              Content strategist
              ┌──────┴──────┐
       Content manager   Content designer
```

In this first structure I've assumed the bare essentials – in what would be the 'Managed' level. A content strategist defines the overall strategy, tone of voice, and governance process, a content designer works alongside design teams to define the content format and structure, and content manager creates and manages web content. In a company that also has content created by marketing, it's prudent for the content manager to work closely with the marketing team on 'shop window' content, and for the content designer to focus on product and purchase content – as they're skilled in creating content for usability.

Centralised

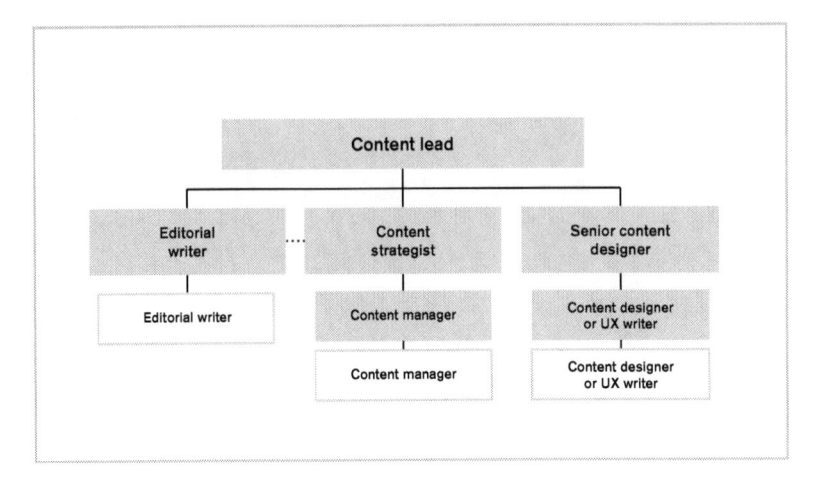

```
                        Content lead
              ┌──────────────┼──────────────┐
         Editorial        Content       Senior content
          writer     ···· strategist       designer
              │              │                │
         Editorial       Content         Content designer
          writer         manager           or UX writer
                            │                │
                        Content         Content designer
                        manager           or UX writer
```

As your team grows you'll create more hierarchy, and add more content managers and designers according to the needs of your brand and product teams. You may also want to include editorial writers who can focus purely on long-form articles such as blog posts. Editorial writers are specialists in creating engaging and shareable rich content. This is probably a similar structure to a 'Centralised' content maturity model.

Embedded

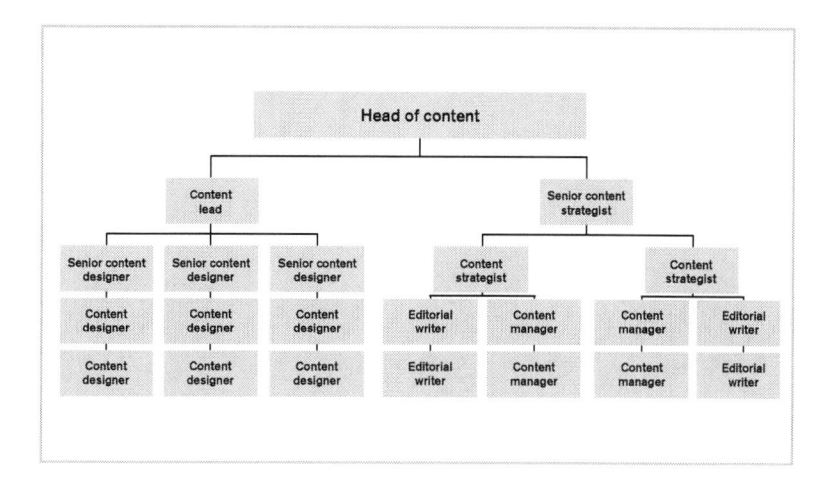

```
                        Head of content
                    ┌──────────┴──────────┐
                Content              Senior content
                 lead                  strategist
          ┌────────┼────────┐      ┌───────┴───────┐
    Senior   Senior   Senior    Content          Content
    content  content  content  strategist       strategist
    designer designer designer  ┌────┴────┐     ┌────┴────┐
       │        │        │   Editorial Content Content Editorial
    Content  Content  Content  writer  manager manager  writer
    designer designer designer   │       │       │       │
       │        │        │   Editorial Content Content Editorial
    Content  Content  Content  writer  manager manager  writer
    designer designer designer
```

In a very mature organisation, there may be a head of content defining overall strategy, guidelines, and governance. But this person's remit is also to oversee capability building, and keep the multiple teams connected, as they're now dispersed around the business. This diagram shows hows you may have groups of content designers, managers or writers, working on different teams, but managed by seniors and leads.

Chapter 7

What are the roles you need in your team?

As you've just seen, in an ideal world you'd have a variety of strategists, writers, designers and pure content managers, led by someone experienced in all areas of content.

In the next few pages I'll explain the different roles, what they do, and how they work.

Content strategists

A true content strategist understands how to take objectives and determine the strategy to achieve the KPIs. They're able to see the 'big picture' and understand how to meet different business requirements.

This person should understand business and managing stakeholders, and be able to work with brand, marketing, product and delivery teams to give content a focus and purpose.

I would always look for someone with an understanding of marketing communications, perhaps with a background in this area. They'll need to understand the role of SEO, social media and CRM (customer relationship marketing), and have the ability to work with personas, carry out content audits and help 'steer the content ship in the right direction'.

At a micro-level they should also be able to make a call on which form the content should take, how to structure it, and be able to oversee the production of it, whether it's words, voice, video or another format.

A strategist should be able to turn their hand to problem-solving. Let's look at some examples of strategy work:

Traditional content-planning and creation

I'm going to help explain this role by using an example business issue.

A content strategist has been asked the following question:

> 'How can we use content to give customers a better understanding of how our product benefits them? And what channels should we use?'

This question seems to be based on some amount of customer insight – that people aren't buying the product because they don't understand how it would benefit them. So the purpose for the content is to better educate or inspire these customers to take further action.

A content strategist would start by understanding who the target audience is and what motivates them to buy. They would probably also want to understand these users' online habits and behaviours – where do they shop, what social channels do they use? Then they'd look at the current content (the content that seemingly isn't doing its job well enough).

They'd define the opportunities to improve or replace existing content, but they wouldn't stop there. They'd understand which channels to use (website, customer communications, or social media, for example) to convey the message, and then identify the most appropriate content type to educate users.

Content strategy goes much further than just improving and creating content. In short it will look at:

- Hosting (where, URLs, page names)
- The content itself (content type, format, frequency)
- Information architecture and navigation
- Traffic driving activity - how people will find your content
- Seeding - can your content be pushed into other channels such as social
- SEO - meta data or tags
- How all your content hangs together across the customer journey
- Creating style guides and tone of voice guidelines
- Call to action and navigation labels
- How the digital product could evolve over time from a content perspective

The strategy they define will set success measures to check their content is working; for example click-through rates, page views, dwell times, or shares. All of these will help show whether customers are engaging with the content (i.e., reading it or being inspired to share it), if targets aren't met the strategist should have a plan for further improvements. It also puts a value onto content – something that can be missing from a lot of businesses. Driving improvements to web journeys may be an obvious output, but defining this with numbers is often a necessary task in order to get 'buy-in' to content from senior stakeholders.

Micro-content strategy

I'm calling this example micro-strategy as this is on a smaller scale, but it's by no means less important.

Here's an example where a strategist might have been asked something like:

> 'On this web page how can we make sure our users know where to go next?'

As above, firstly the strategist will understand who the users are, and the context of this content in their journey. They'll also want to know where the users have come from and what their needs are on the page identified. Are their needs being met?

In this case the purpose of your content is direction, i.e. what you want your user to do next.

Directional content can take many forms, and of course it might just be a case of simply tweaking micro-copy or button labels. But before taking action there are some other fundamental questions that need to be answered by a strategist.

In content strategy there are often overlaps with other disciplines, so sometimes this work's done in tandem with other people, such as designers.

This strategy will include:

- Message hierarchy - is the page providing the right information in the right order?
- Site information architecture - is the page sitting in the appropriate place on site?
- Navigation - are the call to actions (CTAs) clearly defined?
- Wording - is it crystal clear and compelling?

The measures set out under this strategy will may be targets such as click-throughs, successful completion of task, page views, or even sales. It depends what you want your user to do.

This kind of strategy work can also be done by a content designer (a content designer should have as much strategy skill as they have writing skill).

Content site strategy

My third example is a meaty one. In this example the strategist has been asked to 'bring three sites together into one'.

There may be some objectives here that involve reduction of hosting costs, improvements in efficiency (as content will only need to be managed in one place) and improvements in search visibility.

The strategy here has a broader purpose, as it will need to outline:

- Platform - where will the content be hosted and served?
- CMS - are you using the most appropriate one?
- Management (or roles and responsibilities)- who will own, create and edit/publish content?
- Site information architecture
- URLs and meta data - how can you ensure the migration doesn't have a negative impact on search visibility?
- Navigation
- Build roll-out and phasing of the content migration
- User needs - how do we maintain necessary content and ditch anything not meeting our users needs?
- Evolution - how could the content change or grow over time?

The first step would be to understand who each site is used by and why. The analytics would be used to look at current user journeys through the site. The content expert might work with a researcher to look at typical site users (personas) and the needs those users have. Once all the user needs are identified the strategist would carry out a content audit. The purpose of the content audit would be to have a full list of all content in order to understand whether we're serving those needs. What should be kept and what can be lost in the migration? And of what we do keep, how should it be optimised?

They would then work to define what the architecture and structure should look like. This might be through research from existing site users, or by a method such as card sorting (seeing how customers would categorise the information themselves).

Throughout migration the above points will be considered and you may have targets such as successful completion of task (which can be measured through usability-testing or tree-testing), search ranking targets, or migration deadlines. With this strategic kind of project you may also have longer term targets that can't be realised immediately, but will be ultimately achieved – such as resource and hosting cost-savings.

This sort of meaty project is also likely to be done in collaboration with research, design and developers.

Overall your strategist can take a problem and identify how to solve it, which could be in one stage, or via a longer term plan.

They can communicate the steps needed to be taken, and the order they need to be delivered in.

Example job description for a content strategist

What you'll be doing:
Working alongside digital, marketing and product teams, you'll be responsible for determining content guidelines, tools, processes, and the overall direction of content.

You'll create and own tone of voice guidelines, determine the longer term strategy for content in your company, and set up the process for creating and publishing content.

You might be involved in resource-planning or procurement of content services, such as tools or agencies.

On a day to day basis you'll be consulting with other functions, such as user experience design teams or marketing, to help them understand how to achieve their goals through content, and how content can help shape your product or services.

Required skills and experience:
- A background in content production and communication
- Working knowledge of digital marketing such as social media and PPC
- Experience in creating tone of voice guidelines
- Experience of collaborating with research and design teams
- Working knowledge of CMS
- Experience of user-centred design processes
- Experience of suggesting A/B or multivariate tests

Possible career progression:
Content lead or head of content

Content designers (and UX writers)

It's tricky to talk about content designers without throwing in the UX writer (user experience writer) term.

In my simple brain, the term UX writer is too limited. Google describes a UX writer role as 'crafting copy that helps users complete the task at hand, whilst Dropbox says 'create copy that's straightforward, helpful and human.'

I think that's great. But I prefer people working on product content to be more involved than just providing words. Having 'a writer' encourages design teams to put in placeholder copy and 'throw it over the fence' to the UX writer later on. Indeed that's probably why many businesses haven't seen the value of this role and just let the UX or UI designers throw in the copy, perhaps sometimes even the marketing team.

A content designer gets involved much earlier in a project (whether it's a transactional journey, or a web redesign project).

The value they bring is helping to determine the format, structure, and style of content. As I said before, they may decide the content doesn't even need to be words, or it might even be voice UI. There is still a value for UX writing, and if you have the luxury of splitting these out and having both roles then I wouldn't stop you!

A content designer has a sound understanding of usability; they will test any content created and adjust it accordingly. They also need to write for accessibility, ensuring content can be understood by the majority of users.

They also need a holistic view of the journey (or experience) provided to a customer, so they can understand how their content works across channels. What has a customer seen before the reach the content they're viewing online? What will they see next? How does it all hang together? Context and consistency are underrated principles – they are the difference between designing a thing, and creating an experience.

They also need to gain a technical understanding of what's being built, to understand all the requirements and restrictions which could have a significant impact on the content that needs to be created. This will enable them to craft help and error messages (including field validations).

It's also a great idea to have a content designer involved in a web redesign project – they have an understanding of how content should be grouped, categorised, and labelled, as well as a view on how the broader brand communication strategy can impact how the content is structured.

Ellen de Vries describes a content designer existing as:

> "A form of interdisciplinary collaboration with the design / UX and developer team (and any other practitioners)."

She says:

> "In an ideal world we would all work together at the same time, and the role of the content designer is to act as a kind of ambassador for the brand language, the messaging and the brand's cultural properties. They are also in charge of crafting the information to suit the design, considering things like page hierarchy, character counts, and how the content might need to change over time."

While UX writers work alongside research and design to make sure the copy in prototypes or on sites is as good as it can be, content designers do a little more than just the writing.

As I mentioned before, a content designer wants to understand how best to design the content for a users needs – from where it appears on the page, to the format it takes. They'll need to know the brand strategy too – for example a chatbot or voice UI might not be an appropriate way for your brand to sell a product. It just might not be right for your target customers.

The content designer can also carry out content audits, run IA projects (deciding how content is structured and organised in a web site). At a micro-level, they can create templates for pages and articles, and help decide how to format content within a page (for example would a comparison table be more appropriate than paragraphs of text?).

There are many good reasons to have a UX writer or content designer onboard, not least:

Efficiency

It's often the case in some companies that a web prototype will be built, then 'thrown over the fence' for someone to 'check the words.'

But when content designers or writers collaborate with design in prototyping, and help with the process, your usability testing gets more realistic results. Test participants are exposed to something quite close to the end copy. This is important because whether you like it or not, even if you're only testing functionality, users will get hung up on the words they see before them. The benefit of being able to tweak these words in the context they'll be seen in, and amend accordingly as you test, is immense.

Here's an example:

> 'At a usability testing session, it's observed that the first participant of the day gets stuck understanding a sentence. The researchers note this down to take back to the office for the UX writer to look at. The third and fourth participants also struggle with the same sentence. It takes the fourth one ages to get past it, and they run out of time to finish the usability test task.'

If the content expert had been present at the testing session, they could have tweaked the sentence after the third participant, and seen whether this helped. It's likely it would have done, and the fourth user would have been able to complete the test.

It's not only testing that's more efficient, it's the project design too.

Here's another example:

> 'A designer has created a wireframe for a customer journey and put in holding copy. The design has been approved by the key stakeholder. It's then sent to the content designer to fine tune the wording. The content designer takes one look and realises that the flow of the questions isn't natural, and that two of the fields could be merged into one to capture the information that's needed. They have to go back to the designer to make these changes, who doesn't want to as the work has already been provisionally approved, and they'll now have to go back through the process.'

In this scenario, if the content designer and UI designer had worked together on a wireframe, the journey would have satisfied everyone and would not need to be revised after the stakeholder had seen it.

These are just two small examples, so you can see how over time the inefficiencies mount up without content designer or UX writer involvement on a project.

Context

As I mentioned before, a content expert is well-versed in the brand's tone of voice, since they write for other areas of the web. They're also close to the rest of the customer experience of the brand, as they often work alongside documentation, product and sales teams. content designers and UX writers understand the importance of context, so even if they don't have this knowledge, they'll be sure to gain it before they start work on a web journey to make sure the content's consistent across channels. With this broad knowledge they can identify the parts of the journey that might be incorrect – and work to ensure they're put right. They'll also link in with other content producers in the business, to ensure cross-channel consistency of message and tone. For example, if a customer log in process is changed, a content designer would think about the email the customer receives inviting him to log in, and would check that the email instructions were updated to reflect the new process.

Design for content

The most important benefit of content and design working hand in hand, is that they'll create pages where the design is appropriate to the copy and the copy complements the design. If you're still using the 'design first, words later' approach then you may be familiar with your content designer questioning the logic of a copy box where content might not be needed, or complaining about character restrictions in component modules. You might also have heard designers complaining that the copywriter's written too much copy for the space provided! Working together to design for the content eliminates these issues.

A/B testing

Creating a user journey is just the beginning. That journey will be rigorously measured and optimised over time.

Content and copy is vital in helping to optimise user journeys. Craig Sullivan from Optimise or Die says:

> "I estimate that at least 60% of my tests got their main lift from optimising the words, the button copy, the headlines, the text decoration, the layout, the scan-ability, readability, comprehension and simplicity of TEXT."

That means much of your successful optimisation can be delivered by a content designer simply making your content more compelling, or easier to understand.

Creating best practice

Setting out 'best practice' guidance based on tried and tested content is really valuable. If design decisions are documented by your UX writers or content designers, then these can be shared across members of different project or scrum teams.

Can writers move from long-form to short-form?

In some organisations the content function is there simply to serve marketing and brand teams, either by creating long-form copy or managing the brochureware page content. But those organisations are missing a trick by only thinking of content as editorial content.

It does take a certain skill to switch from rich long-form content production into writing short-form micro-copy. Editorial and marketing content is often driven by business needs, but functional and instructional copy should be very driven by user-needs, so it's a shift in mindset.

A UX writer or content designer needs to empathise with customers first (whilst at the same time deliver the needs of the business). They'll have an understanding of usability and will always be striving to crystalise their copy to something clear, and crisp, yet friendly.

They'll also have to think about regulations and legal requirements, have product knowledge, understand brand and product strategy, and accessibility needs.

They need thick skin, as user testing, listening to what users don't like or understand, and iterating copy is a vital part of this person's role. They'll also often have to deal with tricky stakeholders. Another part of this role is working with conversion experts to create content for A/B tests and optimisation.

Functional copy needs to be minimal. Especially when you think about how users consume copy on small screens! UX writing is about micro-copy (and micro-detail – you need to understand each and every possible user path!). Sometimes you may need to communicate with just a word or two (or even no words at all). Reducing something complicated to a small explanation that's accessible to all requires a different skill-set to writing a long-form article.

An intuitive web journey is one where the copy doesn't need to be thought about by the user – they shouldn't even notice the words, because they blend seamlessly into the overall experience. In fact one of my favourite quotes on UX writing is by Ben Barone-Nugent who said:

> "I dream of interfaces where all my words can cascade away during the journey."

Content within a journey should be so simple to understand that it's almost subconscious to the user. This is different task to long-form writing, when you want a user to engage and take in every word.

Moving from long-form to short-form can be done, but it's not for everyone, which is why I'm always very wary of the attitude that anyone can put the content into a web journey. You'll reap so much benefit from recruiting content designers, who can take your digital content to the next level.

Example job description for a content designer

What you'll be doing:

Working alongside research and design, you'll be responsible for structuring and designing user-centred content, to help create a seamless online experience.

You'll be collaborating with other disciplines on a daily basis – understanding issues and opportunities, mapping out design concepts, and testing and iterating accordingly. When writing content, you'll also be working alongside relevant subject matter experts and stakeholders, and managing approval processes.

Part of this role means optimising live content – working alongside analysts, product managers and designers to spot issues to improve conversion through improved usability.

Required skills and experience:
- Excellent communication skills and proven copywriting experience
- Experience of using tools such as journey maps and empathy maps
- Experience of collaborating with UI design or product designers
- Experience of prototyping tools
- Experience of attending and assisting in research and usability testing
- Experience of suggesting A/B or multivariate tests

Possible career progression:
Content lead or head of content

Example job description for a UX writer

What you'll be doing:
You'll work alongside design teams to create user-centred copy for web journeys.

On a day to day basis this means collaborating on wire-frames to produce prototypes, attending usability testing and iterating accordingly.

You might also be involved in helping to produce the tone of voice guidelines and messaging frameworks for the brand.

As part of this role you'll work with analysts and designers to identify opportunities to improve the copy on existing live journeys as well.

Required skills and experience:
- Excellent communication skills and proven copywriting experience
- Experience of working in collaborative teams, particularly with designers
- Experience of prototyping tools
- Experience of attending (or running) usability tests

Possible career progression:
- Content designer or senior content designer

Content managers

Traditional content managers have a good working technical knowledge of content management systems and publishing processes. They're the ones who'll roll up their sleeves to build, test and deliver pages that look great and flow well. They'll be working closely with brand and marketing to help create their content and ensure consistency but they'll also work closely with front-end developers.

Bug fixing is also a part of this role so this person will be good at trouble-shooting and spotting issues before the users do!

Content managers also have to work well under pressure and be able to flip from one content management system to another, so adaptability is a key skill for this person!

Your content manager will need to:

- Be self-organising and unflustered under pressure
- Have unwavering attention to detail
- Have experience of various content management systems and basic knowledge of html
- Have SEO knowledge
- Understand the importance of meta data and alt-tags
- Be able to write clear and concise copy – long form or short form
- Be able to edit and proof-read
- Be able to communicate effectively with dev ops teams
- Have excellent stakeholder management skills

A content manager will need to have strong time management and prioritisation skills. They'll often have multiple (and sometimes conflicting) demands coming at them from different stakeholders, so they'll need to be able to respond accordingly.

As your content manager is your last checkpoint before publishing, they should be conscientious and diligent – almost obsessive about the details.

Example job description for a content manager

What this role will be doing:
Working alongside brand, marketing, product, and content teams to produce and publish content to the website.

You'll be expected to be the day-to-day guardian of the website, owning the content and the publication process.

You'll typically be involved in managing campaign or product updates, ensuring any new content is produced to a high quality and complies with any necessary regulatory guidelines, and that it's optimised for search. This may mean managing the approval workflow prior to publishing.

You'll also ensure live content is kept up to date, and that the site is functioning correctly, with any identified issues being fixed accordingly. This means you'll work closely with developers.

Required skills & experience:
- Experience with various content management systems (CMS) and willingness to learn new ones
- Knowledge and experience of SEO best practice
- Good communication skills and experience of managing stakeholders
- Proven experience in content production (articles, short-form copy and video)
- Basic knowledge of code such as html

Possible career progression:
Content strategist

Content lead

As well as having a broad understanding of the above three roles (ideally have done all three roles at some point themselves), your content lead will be responsible for most (if not all) of the following:

Creating principles

Setting the guiding principles of content – style guides, tone of voice, best practice etc. This is the cornerstone of consistent content creation across the team. The leader needs to set a level of quality for the team to work to and ensure that the team are following the principles. Your principles might be practical ones, such as:

- Knowing your customers well
- Understanding the purpose for your work
- Setting success measures
- Being clear on the context
- Following a sign-off process

Or they might be more abstract team principles, such as:

- Ensure content is visible and coordinated
- Create content that's accessible, clear and in plain English
- Only produce content you'd be proud to put in your portfolio
- Never stop learning and sharing knowledge

They might even decide on a combination. It's the lead's role to establish what guidance and motivation the team needs in order to create the best content they can.

Setting a strategy and objectives

The content lead understands the business needs and ensures the content team are working to meet these objectives. This might be by creating a team vision, or it might be by producing team objectives that are stronger together as a whole. Either way, the team should also be structured by the lead in such a way that enables the business to achieve its goals, and enables each team member to achieve his or her goals. I'll talk more about objective-setting later.

Resourcing and recruitment

Leading a team means being able to identify which skills are needed and where, then recruiting to meet the skill gaps. The lead therefore needs to be diplomatic, confident, and able to influence. They'll have to select personalities that complement each other, whilst remaining diverse enough to provide depth and experience across the team.

Mentoring & growing capability

It's hard to ensure the team members are always learning and growing, and their training needs being met. But the leader will need to facilitate this and decide where to invest training budgets.

Ensuring quality

Principles and guidelines are one thing, but ensuring they're followed isn't always easy. What the team lead should do is have visibility of the team's work to ensure quality and consistency – but without micromanaging and staring over shoulders! Building mutual trust is the only way to find this balance. Good management is about creating an environment for team members to do their work at the best of their ability, giving them space to grow and thrive. To me it's more about mentoring and leading than 'managing'.

Encouraging collaboration

Your content lead not only needs to ensure teams are working together and that content is an integral part of the research and design process, but they need to also make sure the content team is collaborating internally too, and sharing best practice. They might decide to set up a community of practice to help with this if the team is spread out across different departments. At these sessions members can talk about work they're doing, but also discuss projects where they might need help with or a second opinion. They might also want to share learnings they've gained by doing something in a particular way.

Building advocacy

Passion is so important for a leader, as they'll need to build external knowledge of content principles, and train wider the business areas, particularly in organisations where there is still scepticism of content and even hostility towards it. Demonstrating the value of content is the only way to secure more investment in this area. A team leader is happy to shout about successes, and share knowledge with those who do (and sometimes don't) want to learn more.

Attracting talent

The content team lead should also produce thought leadership articles, or take part in public speaking or social media to build their external perception and attract talent. If it can be seen that the team are doing amazing work, and really driving content-led production, it'll be much easier to recruit new, and talented team members.

A note on content ops

It's been recognised that some areas of content – such as creating and managing style guides, and creating processes – are more operational, and as such could benefit from someone whose role is purely content ops.

I think if you have the luxury of budget, this could be a great area to develop, as the operational side does rely on someone process-driven with meticulous attention to detail. For now I have included op-related tasks in the team lead function, as I see them as creating best practice for the rest of the team to follow,

But breaking them off to another person and allowing the leader to focus on growing, mentoring and recruiting could be a great option.

Example job description for a content lead

What this role will be doing:
Managing a team of content experts, you'll be responsible for the content created for brand and product.

You'll determine the overall strategy for content in your company, and decide how your team should be structured to meet the demands of the business.

You'll oversee the creation of process, and ensure quality and governance of content.

You will be expected to lead, mentor, and help your team members be the best that they can be.

This involves recognising development and training needs, and encouraging feedback through crit sessions.

You'll also ensure your team is represented outwardly – by working alongside key stakeholders to grow the profile of content and encourage cross-team collaboration.

Required skills and experience:
- A background in content production and communication
- Working knowledge of digital marketing or product development
- Experience in creating tone of voice guidelines and content strategy
- Experience of stakeholder management at a senior level
- Working knowledge of CMS
- Experience of running content workshops or collaborating with UX and design teams
- Experience of process mapping and defining new process
- Experience of managing content experts

Chapter 8

Behaviours to look for

Of course team members' skills aren't mutually exclusive, there's often overlap, and I encourage my team to 'flex their content muscles' across other skill areas! But it's rarely the case you'll find one person who can cover all three areas effectively. A good content team will have much more to offer than words alone. Build a well-rounded team that can cover all skill areas, and they'll serve your digital team and your business impeccably.

I believe there are some team behaviours that every team member needs. Decide which behaviours are important to you and look for candidates that display them. If you compromise on any of these behaviours you could find the rest of team picking up the slack, which is never good for morale.

A good team should learn from each other, and be happy to support each other. But I also look for:

Storytelling

A content specialist should be able to tell a story – whether with words, images or video.

Telling a story relies on finding a way to resonate with the target audience effectively, and finding an emotional hook.

A storyteller can turn evidence into a compelling case to influence key stakeholders. They can take dry information and make it interesting to customers. And best of all, they can add often bring a different perspective to projects and provide insights in a way that's useful and valuable to any proejct team.

Passion

I aim to have a team who have a desire to learn and improve. This doesn't necessarily mean they want to take on lots more work over and above their day job. What it means is that they have enough passion about content to want to look externally for inspiration, and learn from other people or industries. Continual self-improvement will benefit the whole team.

This means they'll probably write blog posts, follow other professionals on Twitter, read books and have view on how to create enthusiasm and advocacy around the business. In an interview they'd be able to tell you which brands they admire and why, or who they aspire to.

Those who are passionate about their job will always go the extra mile to do what they believe is right for the user.

Good communication

Of course content is all about communication. But I look for people who can communicate verbally as well as digitally. This is really important for stakeholder management, but also for sharing knowledge at team meetings or wider business meetings.

Being a good communicator is vital for working in agile scrum teams as you must be able to voice concerns and be open with the rest of the working team. It's also vital for collaborating and pair-writing with other areas of the business.

Problem-solving

Much of a content expert's role is to decide how we can meet a user need through content. Often this involves taking a complex amount of information (back-end process, context, product information, regulatory restrictions etc) and coming up with a solution that presents something simple and intuitive to a user. Hiding the complexity from the customer is the objective here. Good content experts welcome the challenge of a tricky problem.

Conscientiousness

Attention to detail is crucial for a content producer, but I also want my team to take pride in their work. I always look for people that want to make something that's high quality, technically correct and engaging, as some of our digital content may have to serve its purpose for many years to come.

Curiosity

A good content expert wants to research and understand more about the target users. Constantly questioning why they're doing something is good. An enquiring mind leads to questions, and questions lead us to the core problem we're trying to solve – which is often different to what we first thought it was.

Chapter 9

Recruiting your team

Creating job roles and descriptions is the easy bit! Finding appropriate candidates for these kind of specialist roles can prove somewhat more difficult. Knowing how to attract, and retain talent is more apparent in some organisations than others. For those that have grown their external profile, and mixed with the local digital communities, the task will be simpler. Twitter is a great place to start, as is Medium. Lots of cities have their own local digital recruitment sites, and if you have the time, looking up content experts on Linked In and contacting them directly can be fruitful.

If you have the budget to use a recruitment agency, I'd recommend going to a digital specialist agency. Using a specialist means you'll be more likely to find people for roles such as UX writer or content designer which are relatively new roles to the digital industry.

Twitter also works well as a way of attracting like-minded individuals, especially if you have some thought-provoking content to share that provides examples of the kind of work your team are doing.

Screening CVs

Content specialists don't always come from journalism backgrounds. I know a great content designer from a design background who realised that he was more interested in the communication side of design and veered towards content. I know content specialists who have come from strategic marketing backgrounds, and some who have been in UX. Passion, competence, and experience have always trumped educational achievements for me. So when I look at CVs I focus mainly on the last two positions (and how relevant they are), and the stated areas of interest. Having an interest in creating user-centred design is a must, as is a desire to turn complex information into something simple.

As you're recruiting a content specialist, the CV should reflect key content principles – it should be easy to read and navigate with no large chunks of copy, be written in plain English without jargon, and obviously have no typos or grammatical errors. It should be well laid out, and simply designed.

For me it's a must for a content specialists to have a portfolio site, where I can view samples of their work, and read their blog posts. If a candidate was unable to provide work samples that would be a deal-breaker for me.

The interview process

If a candidate's experience looks relevant, their work is of a good standard and they're showing enthusiasm for the role, your next step is to meet them. Each company's recruitment process works differently, but I like to have a phone conversation first, to tell the candidate more about the role. It's also a good opportunity to sound them out about why they're job hunting. Then I'll invite them in for a more formal chat. Depending on how many candidates I have, and how much knowledge I have about them, I might extend the face to face interview to include some exercises as well.

Phone call

A phone call saves the time and expense of the candidate having to travel to you, but provides a means to gauge interest and experience. This is just an informal half-hour chat. By the end of it your candidate should have a better idea of what your roles is and whether they're interested, and you should have a better idea of them.

Do you get the impression they're particularly interested in your role, or just desperate to leave the one they are in?

I will also explain how my team is set up, the kind of projects they'd be working on, and what our working culture is like.

I would expect the candidate to have one or two initial questions about the role. Take notes during the call, and keep them with the candidate's CV, so you can refer back to them later.

Face to face

This is a chance to put a face to a name and ask a few more questions. These should be more competency-based, you want to get an idea of how capable the candidate is, but also how passionate they are about content and customers. Here are a few example questions you could think about asking, depending on the roles you're recruiting for:

Questions to ask potential content managers

Which content management systems do you feel most comfortable using?
Don't rule out someone who hasn't used your particular CMS. Many systems have similarities. Aptitude is more important – if someone is a fast learner it might not be a problem that they haven't used it before. What I'm interested in here is how technical they are, and whether they have the confidence to find their way around a new system. If they've used two or three different systems easily then they're probably fairly adaptable.

Do you have an example of a time when you've had to deal with a very tricky stakeholder? What happened? What did you learn from the experience?
It doesn't matter in this scenario whether something went well or badly. We often learn more when something's gone wrong.

Content managers in particular have to accommodate opinions on copy from many stakeholders and decide which ones to act upon, if any. I'm interested in how a content manager copes with feedback and influences others.

Sarah Richards puts it more succinctly:

> *"I ask 'How do you handle tricky conversations?' I think a large percentage of our work as content people is to convince others why it is so important. Content still isn't valued in a lot of organisations. I want to know that my candidates know their skill inside and out so they have to be able to efficiently communicate content decisions"*

Is there a time when you've had to deal with conflicting priorities? How did you manage the situation?
The art of priority management is communication, an essential skill for a content expert!

Questions to ask potential content strategists

When being briefed to produce a content strategy, what are the key questions you'd want answered?
I'm looking for someone who understands the needs to align business and communication objectives with the audience needs. Someone who knows that in order to communicate with an audience appropriately they need to understand as much as they can about that audience. I'm also

looking for someone who considers the practicalities and constraints, but can (when needed) think without boundaries. Creativity is a key skill for a content strategist.

Do you have an example of a time when you've developed a new process or system to improve ways of working?
Strategists needs to think as much about process as about content. They'll be the person defining or championing a publication process, or collating stakeholder feedback. As a strategist has the bigger picture, they should be able to spot opportunities for improvement and be able to work with other functions such as marketing or brand, to influence change.

Which brands do you admire for their content strategy?
A strategist should understand the value of cross-channel consistency and I would expect them to have a view of brands that provide a truly great omni-channel customer experience through content.

Questions to ask potential content designers

What is the best collaborative project you've worked on? What did you enjoy about it? What did you learn from it?
I have found that you learn the most from working with other disciplines and stakeholders. As this is a crucial part of the role, I'll be listening to hear about what the candidate takes from collaboration and whether they enjoy it or not!

Do you have an example of a time when you've had to deal with a very tricky stakeholder? What happened? What did you learn from the experience?
This is probably one of the most important things, Sara Culver agrees, she says:

> "I always try to get a sense of how candidates deal with stakeholder alignment, since content always has many stakeholders. Everyone can write, and almost everyone has opinions about words, so content folks will often be getting lots of input from different directions and still be ultimately responsible for the words that end up in the product and whether they succeed or fail. I don't think there's one right way to approach these situations, but I'm on the lookout for folks who have thoughtful answers, and who know that engaging stakeholders successfully is part of the job. There's no "my way or the highway" in content."

This is another essential for John Saito, who says:

> "I look for thoughtful writers who can easily explain their writing decisions to others, just as a designer should be able to explain their rationale for every little design detail."

Outside of your job, what do you do to further your digital content expertise?
In my experience, employees who show passion for their work will always be looking at external influences to further their knowledge. This doesn't need to be an all-consuming passion, but someone who reads blogs, attends conferences or even just speaks to other content designers displays a willingness for self-improvement.

Which brands do you admire for their copywriting?
I would expect a content designer to have a view of brands who do content design well, and why they think that.

How do you approach writing copy for a web journey?
This question will tell me how proactive a candidate is. Are they just waiting until they are asked for some copy creation, or are they throwing themselves into a project in the discovery phase to get the research and insights they need to attain the user needs? Are they used to testing their copy through usability sessions, or just pushing copy live and hoping for the best?

Exercises

I don't think there's any need to overdo the exercises if you've already seen evidence that someone can create great work. But if you have a few candidates to choose between, an exercise or two can sometimes help you to rank them.

If I am hiring for a more junior role, I'll be checking that the candidate can write under pressure. In this case I may simply provide a complex piece of information and give them fifteen minutes to re-write it into plain English for a website.

Of course I will leave the room at this point, it's not an exam!

For a more senior role I might set a small piece of pre-work. For a content strategist I would create a mini brief; identify a problem (or an opportunity) your business has, and ask the candidate to present back for ten minutes (at their interview) on how they would use content to address this.

Keep your brief very high level; part of this challenge is to see whether the candidate asks you further questions or acknowledges the lack of research they have when they present back to you. If they do, but they also show their assumptions and rationale for their decisions then that's great.

If I am interviewing a content designer or UX writer my task will be a little different. In this instance I like to take an example of a poor web journey (from the content and UI design point of view), and pose a question. It might be something like 'This page isn't converting very well. Why do you think that is, and what would you do about it?'

I'd be looking for them to make some suggestions on how to improve the content and copy of the page. This type of exercise can be given as pre-work or done as a twenty minute exercise at the interview. I'd also be looking to hear what they considered but maybe didn't go with – and why not.

These are just some examples, as I'm sure you'll find your own way to judge your candidates. If you've whittled it down one or two candidates but you're not sure, also think about personalities – who do you think seems adaptable? They'll need to collaborate well, or sometimes be happy to work alone, not take offence to critique of their work, and be willing to share and learn.

Chapter 10

Effective team management

A team lead's job is to help make their team the very best they can be.

Empowering your team to get on with the job without 'checking their work' is important, but it can sometimes be hard not to project an 'I wouldn't have done it like that' approach. Team members can all have different writing styles, and that's important — it's the little details that bring copy to life, and a small element of personality in writing is a good thing. But at the same time, you need to ensure consistency and best practice. This is even more important if a team can inter-change on projects when necessary.

Guidelines, in the form of tone of voice frameworks or styles guides are important. But there are other things you can also do to ensure you're all working towards the same ambition in a similar way.

Onboarding

When your new team members start, what do you currently provide? A quick health and safety induction, and maybe some stationery and their laptop?

Here are some things you can do to make a new team member feel welcome, but also get them to start thinking about what you're doing as a company, and how they can personally help you get there.

- Provide a brand induction - who are you, where have you come from, and where are you going?
- Provide your team's vision and objectives - what do you stand for?
- Provide something branded if possible that brings who you are to life
- Take the new member for a team lunch/coffee/after work drink
- Set up intro sessions with the key stakeholders they'll be working with and the key members of the wider team (e.g. your design lead or lead researcher
- Provide some 'housekeeping' information — when will they get paid, where do they record holiday, what apps do they need to install?
- Have a first one to one, and find out how they like to communicate, receive feedback, and what they need to do their job effectively

Team vision and goals

With a new team in place, it's important to provide a purpose (over and above delivering the business objectives). Find out what gets your team out of bed in the morning and find out how you can help them to do more of that kind of stuff, and less of the boring stuff!

In order to provide the best service to the business, it's important to have an ambition to drive your strategy. You should make your team feel they play a role in defining this.

Run a session and ask "As a team, what are the three things you want to be famous for?"

This can help you create a vision that drives your team's own personal development goals, as well as meeting the needs of the business.

Here's an example of how one such vision worked in practical terms:

> 'One of the things my team wanted to be known for is being invaluable in prototyping. To be engaged from the start of each project we decided we needed to make a nuisance of ourselves and prove our value – until we were always automatically included. This was a team behaviour we agreed to adapt to meet our goal. We went from being the team that always received a prototype (to put the content into) last, to being the team creating content-led, conversational web journeys, within the space of a year because of our tenacity to achieve our goals.'

Annual planning

Take some time to do some annual resource planning for the year. Look at what's coming up in terms of project delivery, across which months, and see where your bottlenecks are. Could you manage this work within the team, or will you have to supplement with some freelancers?

Find out what your business targets are for each project. These commercial values will be what your team's success will be measured against. Your content should be a key contributor towards the business goals (which of course you'll be aiming to achieve by meeting user needs). These will form the basis for your team's business objectives.

Setting objectives

Now we have identified the business targets, and we know what we want to achieve as a team to establish our value.

This should translate nicely into some individual objectives. I believe objectives should be a combination of business goals (e.g. help achieve x sales, or a x% increase in conversion), team targets (e.g. achieve a ratio between content and design of 1:1 across all project teams), and personal targets (e.g.. Speak at 1 external conference event to gain more speaker experience).

It does depend on the projects that each member is assigned to, and of course on their own personal development targets. And if you set targets for your team to meet, it's important to remember that if you move them onto a different project mid-year, you'll need to revisit their objectives and check whether they're still relevant.

Targets should be measurable, and also achievable. I've learnt not to try to stretch people too far, but a little nudge beyond the comfort zone is good!

From the rest of the year, meetings could take up the following rhythm:

A weekly one to one catch up

This is just to see whether your team member needs any help with anything. It helps to masquerade this meeting as 'going for a coffee', as a walk out of the office can make someone speak more openly about something that's bothering them. It doesn't matter so much if this meeting doesn't happen every week. But as long as they take place most weeks your team member will know you're approachable should they need you. You can use these sessions to see whether there are any issues or blockers to their work, or just to gauge how they're feeling about work – exhausted, excited, or frustrated. Find out what you can do to help.

A monthly one to one

This is a review to check in where your team member is against their objectives. It's also a dip-check on how they're feeling about their career – do they have training needs, are they happy with the projects they're working on? Use their objectives as a discussion guide, but try to focus more on personal development. It's also a good opportunity to ask how fulfilled they're feeling. Is their work challenging them or boring them? Do you need to move them onto a different project to keep them learning? I had a great line manager once who continually pushed me (gently) out of my comfort zone as she knew I could do more. I didn't think I could, but I managed to go way beyond what I ever thought was achievable. When someone helps you fulfil your potential, it feels great and is a big confidence boost.

Bi-annual review

This is a more in-depth one to one. It can be valuable to get anonymous feedback (via a survey) for the team member.
You can get great feedback with just three questions:

- What value has X brought to your project?
- What development areas would you suggest for X to work on?
- Would you recommend X to another project team?

Sometimes you'll have a view of how a team member is performing, and often the feedback will support and validate those views. Personally I always find the development areas for myself very valuable, it teaches me what others perceive to be my weaker points. But bear in mind that some people won't be as willing to hear that they have areas to work on. Try to frame these as opportunities rather than failings. Provide solutions to work on them. For example, if a team member needs to develop their presentation skills, help them find an appropriate workshop to attend and give them a small, achievable target to work towards, such as running a 'lunch and learn' in a safe team environment first. It's a gentle push out of their comfort zone, a step at a time as they gain confidence.

Weekly team review session

I used to run these as a 'round the room' catch up. But the reality is that this offers little value to the team without tangible work examples. I began to emulate the design peer review sessions and asked team members to bring some project work they'd like help with or feedback on. Some team members had come from agile digital teams who were used to crit sessions (critique sessions) so they embraced it. The other members were slow to bring work. While discussing this at a one to one session I discovered they didn't bring work because they didn't think their work was as 'meaty' as some of the content builds we'd been sharing. What they didn't realise was that despite their project being smaller, it was actually very valuable, as they'd started to see some real conversion uplifts simply through iterating wording. Once they realised they had as much to offer as everyone else, their attitude towards these sessions changed.

I therefore realised I perhaps hadn't set the ground rules very well or explained that all work is valid for these sessions, however small.

Team meetings are also a great time to discuss anything affecting the whole team, and how it could be approached together, or ideas you all have for sharing best practice (or new tools or apps you've found that could facilitate this). It's good to do a quick round the room as well, for an update on what everyone is working on, but it shouldn't be the sole focus. And in fact you could do that in a weekly stand up instead.

Ellen de Vries says:

> "Apart from doing the job, content roles have such diversity that cross-pollination of ideas, education and shared curiosity around processes, methodologies (and negotiating organisational infrastructures) are vital."

So use these sessions for discussions, sharing ideas to document process and best-practices, and also just sharing any great external work examples you've seen.

Facilitating reviews

In order to help facilitate review sessions, I took my lead from designers and created some prompt cards. They're designed to challenge the content to ensure key principles have been followed, by posing a question. The principles are:

- Purpose
- Branding
- Format
- Simplicity & clarity
- Quality
- Design & content working together
- Measurement
- User testing
- Customer insight

So for example a question might be "How will your user know what's expected of them?" or "How will you ensure stakeholders that need to be, are comfortable with your content?"

They're open-ended to incite discussion.

This is just one example of how these sessions can be facilitated. Previously I've also posed them as surgeries: "Does anyone have anything they need some help with?" which can be a good way to encourage people to share work and ask for help.

End of year wash-up

Inspired by a blog post by a designer friend Michael Allen, I ran a team review. It seemed to work really well so I'd recommend it. I asked four simple questions:

- What are your expectations of the team
- What are we doing well?
- What isn't working so well?
- What should we be doing more of?

This provided a useful lens to view how your team feel about the work they're doing, and what more you could be doing for them as their leader.

Once you have all the output on Post-its and grouped, you can clearly see the actions you need to take forward into the following year. Some may be in your control, others might rely on you working with other stakeholders, or managing those message upwards. But do summarise your outputs and actions so that your team know you're doing something to improve.

And finally...

I didn't want to make this book a heavy read. But I wanted to drive understanding of what you can achieve by establishing content roles and how you can expect content specialists to benefit your business. Some people will have their own views on what content should, and shouldn't do, and some people will classify content roles differently. Content runs through every aspect of business and service design, so by its nature it's complex and nuanced.

Of course there will always be overlap with design and UX, brand, marketing, PR and legal, and many other areas of your organisation.

The point of a team, is that they're stronger together as a collective, and so even if you start small the sum will be greater than its parts. In fact, even starting with just one content specialist, if currently you have no one at all, will make a difference to your product and how you speak to your customers.

I have many blog posts on digital content writing and strategy, so do have a look on Medium if you'd like to know more. I've also added a reading list at the end if I've whet your appetite to learn more about content.

Contributors and further reading

I'm so grateful for everyone who's helped me in pulling this together. Some very talented folk have lent me their time and opinions. If you'd like to read more about content, I can highly recommend you look up my contributors and their blog posts or books.

Ellen de Vries, content strategist at Clearleft
I recommend her book: Collaborate - bring people together around digital projects, which you can download at:
https://gathercontent.com/books/collaborate

John Saito, UX writer at Dropbox
You can read more of his wisdom at https://medium.com/@jsaito

Martyn Reding, head of user experience at Virgin Atlantic
An amazingly skilled design leader who helped design this book, you can follow Martyn at twitter.com/martynreding

Sara Culver, senior product writer at Slack
You can follow Sara on Twitter @sj_culver

Sarah Richards, content strategist
Sarah's book Content Design is available now on Amazon and a great read for anyone, content expert or otherwise.

Recommended reads:

Nicely Said by Nicole Fenton (New Riders, 2014)

Connected Content by Carrie Hane and Mike Atherton (New Riders, 2017)

Content Design by Sarah Richards (Content Design London, 2017)

Conversational Design by Erica Hall (A Book Apart, 2018)

How to Make Sense of Any Mess by Abby Covert (Abby Covert, 2014)

About the author

Rachel McConnell has spent her career creating strategy and designing content – from brand and marketing communications, through to social media and digital product design.

She's worked on high profile brands such as Deliveroo, Nationwide Building Society, John Lewis, M&S, Tesco, and Flora, which provided some unique insights into how content is (and how it should be) developed in large organisations.

Having also built and managed teams, Rachel's seen the value content experts bring to a brand experience, and is keen to help other companies grow their content capability.

When she isn't writing or thinking about writing, she likes to get outside – to the forest, beach or garden – and spend time with her family.

www.rachelmcconnell.me

ISBN: 9781720128441

Design by Martyn Reding

Imprint: Independently published

First printed November 2018

Copyright © Rachel McConnell 2018

Printed in Poland
by Amazon Fulfillment
Poland Sp. z o.o., Wrocław